NATURE CLUB

P9-DYE-893

NIGHT CREATURES

JOYCE POPE

Illustrated by
TREVE TAMBLIN

Troll Associates

Sandra Hoffmann

Nature Club Notes

Though you may not know it, you are a member of a special club called the Nature Club. To be a member you just have to be interested in living things and want to know more about them.

Members of the Nature Club respect all living things. They look at and observe plants and animals, but do not collect or kill them. If you take a magnifying glass or a bug box with you when you go out, you will be able to see the details of even tiny plants, animals, or fossils. Also, you should always take a notebook and pencil so you can make a drawing of anything you don't know. Don't say "But I can't draw" – even a very simple sketch can help you to identify your discovery later on. There are many books that can help you name the specimens you have found and tell you something about them.

Your bag should also contain a waterproof jacket and something to eat. It is silly to get cold, wet, or hungry when you go out. Always tell your parents or a responsible adult where you are going and what time you are coming back.

It is difficult for most people to see many nocturnal animals. One way to get around this is to visit a zoo that has a night house. This is kept dark during the daytime and lit at night. The animals that live there are active when it is dark. As a result you can go there in the daytime and watch nighttime animals in perfect safety.

Library of Congress Cataloging-in-Publication Data

Pope, Joyce.
Night creatures / by Joyce Pope ; illustrated by Treve Tamblin.
p. cm. — (Nature club)
Includes index.
Summary: Describes the appearance and behavior of such nocturnal animals as the owl, bat, and moth.
ISBN 0-8167-2783-X (lib. bdg.) ISBN 0-8167-2784-8 (pbk.)
1. Nocturnal animals—Juvenile literature. [1. Nocturnal animals.] I. Tamblin, Treve, ill. II. Title. III. Series.
QL755.5.P66 1994
591.51—dc20 91-45171

Published by Troll Associates

Copyright © 1994 by Eagle Books

All rights reserved. No part of this book may be reproduced or utilized in any form or by any means, electronic or mechanical, including photocopying, recording or by any storage and retrieval system, without permission in writing from the Publisher.

Designed by Cooper Wilson, London
Edited by Kate Woodhouse

Printed in the U.S.A.

10 9 8 7 6 5 4 3 2

Contents

Introduction

Human beings, dragonflies and cows are among the many creatures of the daylight. People can lengthen the amount of time that they are active by using artificial light. But once it is dark and they cannot see, there is not much that daytime, or *diurnal*, animals can do, except sleep.

Some animals live in the opposite way. These are the nighttime, or *nocturnal*, creatures. They sleep through the day and come out to hunt and feed once it is dark. Very few animals are equally active during the day and night.

The *environment* is like a factory. It is most productive if it works all the time, with a day shift and a night shift. Many creatures have similar ways of life; only their times of activity are different. Most butterflies feed on nectar during the day; most moths are nocturnal nectar feeders. Hawks hunt small mammals and birds in the light; owls catch similar prey in the dark. You could make a long list of similar pairs of animals.

Some animals have become creatures of the night to avoid being hunted by humans or other enemies. If there is no danger, they may be seen in the daylight.

When the sun goes down, the weather gets cooler and the air becomes moister. This suits many creatures, such as slugs and some kinds of antelopes, better than the heat of the day.

Nighttime Animals

When you look at an animal, you can often make a good guess as to whether it is diurnal or nocturnal. The clues are in the size of its eyes, ears, and whiskers.

Daytime animals usually have moderate-sized eyes, which enable them to discover food or enemies without using their other senses. Nighttime animals often have larger eyes. Usually they cannot see colors, but their eyes are sensitive to very small amounts of light. They are able to make out details of shapes in what to us would be complete darkness.

▼ The fennec fox lives in the deserts of North Africa. It is about the size of a small cat and is active at night. Though it has a good sense of smell and touch, its sense of hearing is probably most important. Its ears enable it to locate prey such as lizards or large insects. It is possible that the fennec fox can hear prey moving underground, because it often digs to get its food.

Many nocturnal animals have a good sense of hearing. Mammals often have huge ears that they can turn toward the direction of a sound. This sound may be a meal – or a hunter! A good sense of smell is also used by many nighttime animals. With this they can find their way or detect food in complete darkness.

A sense of touch is vital to many nighttime animals. Cats, mice and many others have long whiskers, called *vibrissae*, around their noses. Each whisker has a big nerve ending at its base. The animals can feel their way, because the slightest touch alerts them to an obstacle. Most nocturnal animals move quietly. A few that are usually slow moving but well armored, like porcupines, are noisy.

▼ Armadillos hunt for insects, grubs, and other small creatures at night. Their armor protects them against most enemies.

▶ The forest-dwelling tarsier uses its huge eyes and ears to discover the small creatures that are its prey.

Seeing with Ears

Some nighttime animals hear so well that they can find their way and catch their prey using their ears alone. When this idea was first suggested, people laughed, saying it was impossible to "see" with ears. We now know that many kinds of animals have a *sonar* or *echolocation* system, using their ears and voices.

Two kinds of birds, the oilbird and cave swiftlets, use echolocation. Both live in caves, and use the system chiefly when they are flying to or from their roosting places in the dark. The oilbird feeds at night, but uses its eyesight and good sense of smell to find the fruits that it eats. The sonar sounds made by both birds are like the rapid clicking of fingers.

▼ Oilbirds are found in Trinidad and parts of northern South America. The caves in which they live are pitch black, so they navigate to their nesting ledges by echolocation. They feed at night, wrenching oily fruits from forest trees and swallowing them whole.

Bats use echolocation sounds that are far too high-pitched for humans to hear. They make rapid sounds – up to 200 in a second. Because of this, their sonar system is far better than that of birds. They are able to catch insects in flight and avoid obstacles, even on the darkest nights. Though bats are not blind, their eyes are usually very small. You really can say that they see with their ears!

▲ In echolocation, animals hear the echoes of a series of very brief sounds they have made. Their brain works out how long the echo has taken to reach their ears. Since sound travels at a constant speed, it is possible to know how far the sound has traveled.

Badgers and Others

Some animals have become nocturnal so that they can avoid their main enemies – people. In remote parts of Europe, otters play in the daytime. But where there are many people, they do not come out of their *dens* until nightfall. In South America, timid agoutis escape hunters by hiding during the daytime. In North America, muskrats and skunks usually stay hidden while it is light.

▼ The muskrat usually does not leave its burrow before dusk, though during winter it may feed in daylight. It finds its food in or near water. When it is alarmed it can hold its breath underwater for over a quarter of an hour, or it can swim underwater for more than 325 feet (100 m) to make its escape.

◀ Agoutis will only leave their burrows to feed during the daytime if there are no human beings around. In places where they have been hunted, they only come out at night to look for food.

Badgers try to avoid human company altogether. Almost everywhere, badgers stay underground when people are around. You have to wait until after dark to go badger-watching. If the badger suspects that you are there, it will probably prefer to go hungry rather than risk meeting you.

Badgers are well suited to a life after dark. Their eyesight is poor, but their hearing and sense of smell are very good. They mark their territories and trails with a powerful scent from their musk glands. They can then recognize familiar places and things.

In spite of their size, badgers hunt small creatures, particularly rodents and earthworms, which they can catch easily even in the dark. They have few natural enemies, apart from bears and wolves, and they defend themselves strongly if they are attacked.

▼ Badgers are suspicious of people. Most stay in their burrows until after dark, when they emerge to look for food. Baby badgers wrestle with their brothers and sisters and mark all sorts of things with the scent from their musk glands.

Birds at Night

People sometimes talk about "the early bird getting the worm." This means being up very early in the morning, because most birds wake at first light and go to roost as soon as it gets dark. A bird's chief sense is its eyesight.

Most small birds fly at night when they are *migrating*. If they are crossing the sea, they can't stop until they reach land. Once there, they often continue their journey at night. This is safer, because hawks and other enemies cannot see them in the dark.

Shearwaters and some other sea birds are very clumsy on land. They are excellent swimmers, and feed in the sea by day. But they come ashore to nest only at night,

► Wading birds often feed at night because the mud flats where they dig for worms and shells are often covered by the sea during the day.

▼ Shearwaters' legs are good for swimming, but do not support them properly on land. They nest in burrows which they leave only at night. One shearwater remains with the egg or chick while its partner spends several days feeding.

▼ The kiwi uses its sense of smell as it hunts for worms and other small creatures. Unlike other birds, its nostrils are at the tip of its long beak.

or they would be seized by large gulls or other enemies. Birds that have been feeding at sea are called to the right place in the crowded breeding colony by their mates, who have remained in their burrows looking after the eggs and chicks.

Kiwis are also defenseless birds, for they cannot fly. Like nocturnal mammals, but unlike most other birds, they have a good sense of smell. They use this at night to find worms, insects, and berries to eat.

Birds of Twilight and Night

Have you ever seen a whippoorwill? Not many people have, even in areas where they are common. The reason is that they rest during the daytime. Their colors camouflage them so that they look like pieces of dead wood. You could pass very close to one without realizing that it was there. As the sun goes down, these birds, and their tropical relatives the potoos and the frogmouths, wake up to hunt.

▼ Nightjars and their relatives the whippoorwills live in many parts of the world. They catch large insects, such as beetles and moths, in flight.

◀ Frogmouths live in Southeast Asia, Indonesia, and Australia. Though related to the nightjars, they catch big insects and small creatures like mice.

▲ The swift spends much of its life in flight. Birds that do not have to look after eggs or young often sleep in safety on the wing, high in the air. Flocks of them have been seen at heights of over 3,000 feet (923 m).

Their food is mainly large insects. All of these birds have huge mouths with stiff, bristle-like feathers around the edge which help to guide their prey in. Frogmouths catch ground-living insects and even small mammals such as mice, but the other birds mainly prey on flying insects.

Swifts are birds that are often active at night. When other birds have gone to roost, screaming parties of swifts can sometimes be seen. The birds chase each other at high speed. They then spiral up into the sky in what is known as their *vespers flight.* High in the air, they doze while flapping their wings for up to six seconds, then rest for about five seconds, and so on in safety through the night.

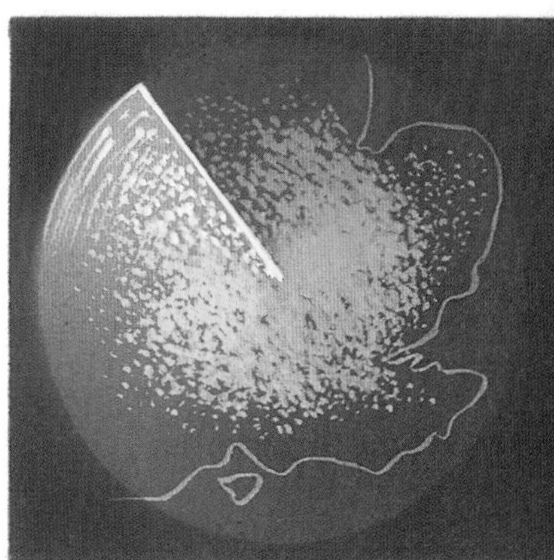

▲ The vespers flight of swifts begins in the late evening. The number of birds may be so great that they can be tracked on radar screens.

Owls, the Nighttime Hunters

Owls are the best-known nocturnal birds. They live in almost all parts of the world, and most of them hunt only at night. By day they are usually unseen, because their feathers match the bark colors of the trees in which they rest.

An owl's feathers are softer than those of other birds. Even its wing feathers are soft, so they don't make any noise when the owl flies. This allows the owl to surprise its prey. It kills its victim with sharp talons. Small prey such as mice are swallowed whole, but larger animals, such as rats, are cut into pieces by the owl's knifelike beak.

Owls find their prey partly by sight – their eyes are said to be able to see about ten times as well as ours in the dark. But when there is no light at all, they must use their ears. Their ears are large, but we cannot see them

▼▲ When there is light, an owl's large, forward looking eyes help it to judge distances. Its feathers do not have hard edges, so the owl flies silently.

◀ An owl's ears are under the loose feathers at the sides of its head. The flaps of skin are movable, which helps the owl to pick up the faintest sounds.

because they are hidden by feathers. Many owls can hear very soft sounds, such as the rustling or squeaking made by a mouse as it runs along the ground. Since each ear is in a slightly different place on the side of the head, the owl can work out the position of its prey and catch it in complete darkness. They do not use echolocation, but they "see" with their ears in a different way.

▲ This tawny owl is pouncing on its prey, which will be killed quickly by the powerful, clawed feet. Tawny owls eat many kinds of small creatures, particularly mice and other rodents.

Bats

Nearly a quarter of all mammals are bats. Most of these night-flying creatures feed on insects. They are the nocturnal equivalent of the insect-eating birds that are busy in the daylight.

In flight, most bats look fluttery and slow compared with birds. Yet their flight is very precise. Echolocation enables them to avoid obstacles and catch insects, even in complete darkness. They can maneuver into tiny roosting spaces in caves or under roofs. They are able to do this because their wings, and the muscles that control them, are very different from those of birds. Birds have two main pairs of muscles to raise and lower their wings; bats have nine.

Some bats have strange-shaped faces and large ears. The purpose of their odd appearance is to help them focus the sounds they produce. Their large ears enable them to catch the echoes – and their prey – very effectively. Bats eat huge numbers of insects, many of which harm us or our crops. A single bat colony in the southern United States is thought to eat about 6,700 tons (6,100 metric tons) of insects each summer.

Some bats that live in warm parts of the world feed on fruit and nectar. A few hunt small creatures such as mice, lizards, or fish, and at least one type eats frogs. Three kinds of bats in South America take blood from birds and mammals.

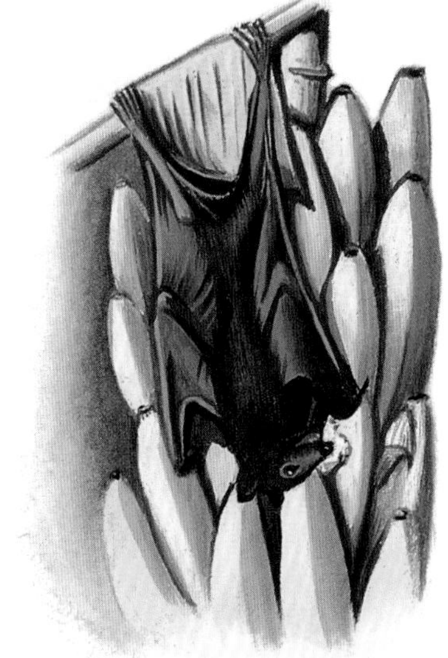

▲ Bats need to find safe places to hide, because once they are resting they often become *torpid*. This saves energy, but they may not be able to wake up quickly if an enemy appears.

► Bats' wings are made of fine, soft skin that is stretched on each side of the body between the hind leg and the long arm and finger bones.

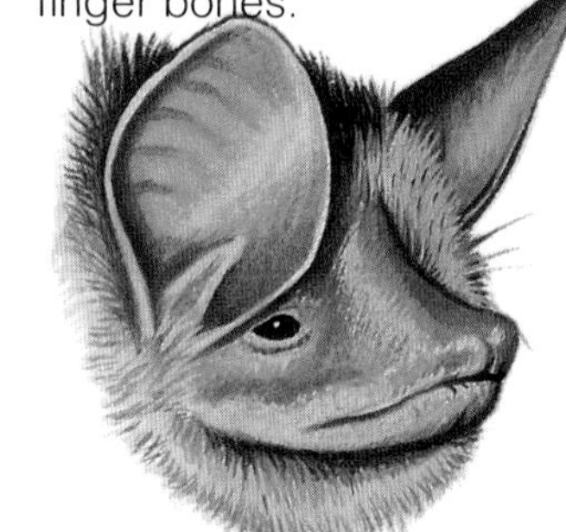

Noctule

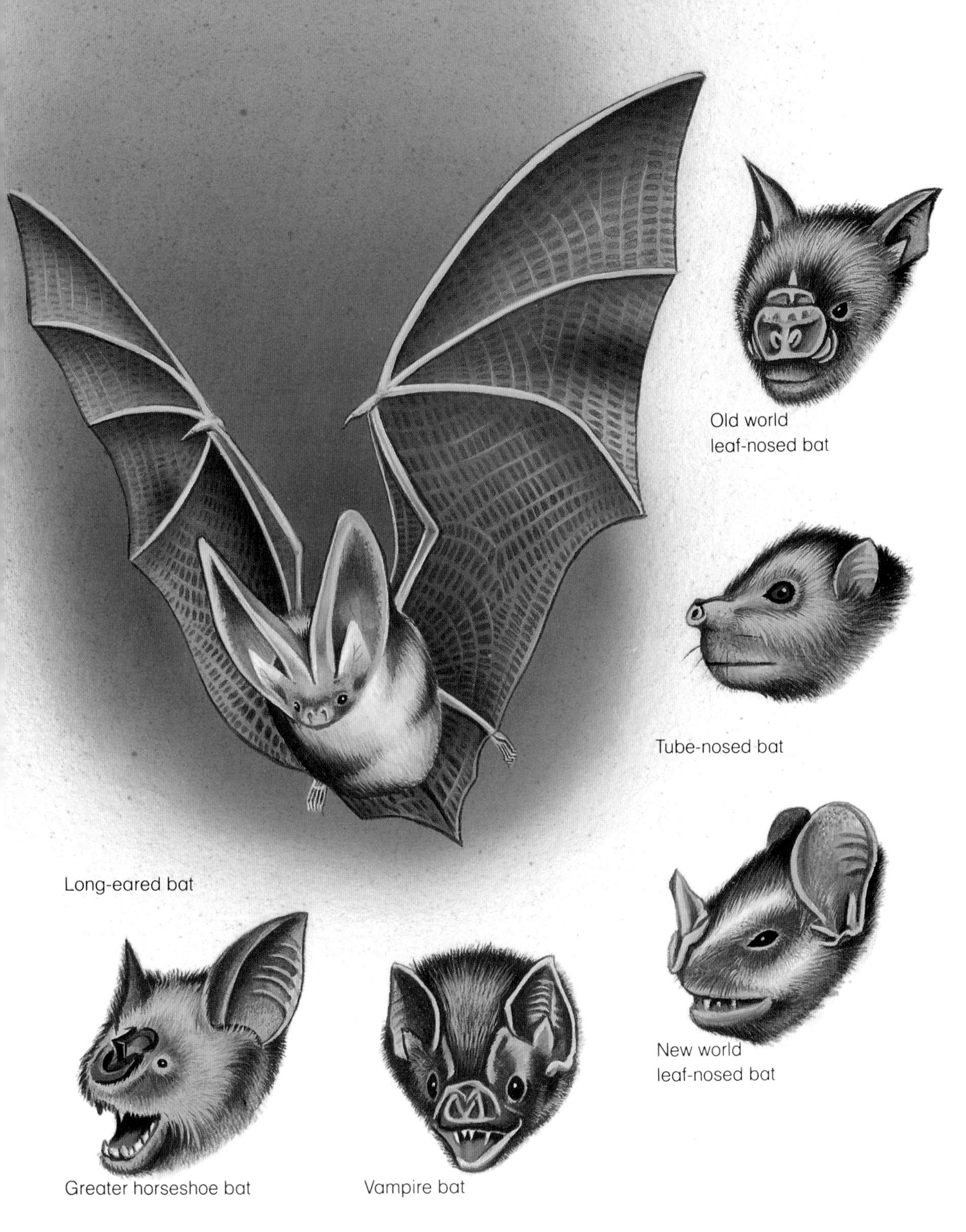
Old world
leaf-nosed bat
Tube-nosed bat
Long-eared bat
New world
leaf-nosed bat
Greater horseshoe bat
Vampire bat

Moths

As light fades, it becomes more difficult to see colors clearly. Because of this, nighttime animals are usually dull-colored. This is true even when their close relatives, active in daylight, are brightly colored. During the daytime, moths have to hide in the open. They cannot make burrows and holes, and many have broad, fragile wings that could be damaged if they were to crawl into crevices. So most of them sit on the trunks and branches of trees. The dull browns and grays of their wings match their background so perfectly that we cannot see them. More importantly, their usual enemies, birds and other insect-eaters, cannot see them either.

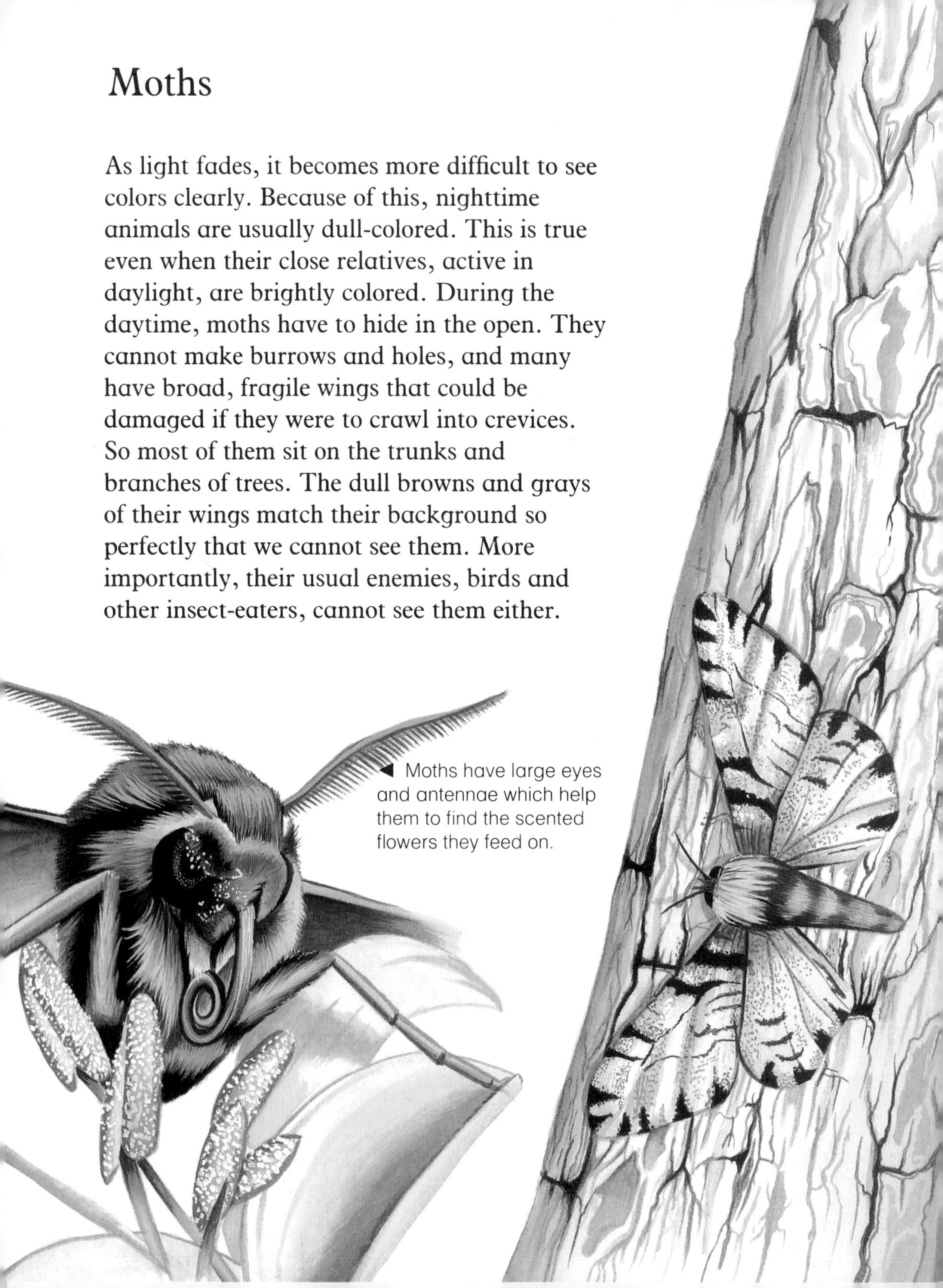

◀ Moths have large eyes and antennae which help them to find the scented flowers they feed on.

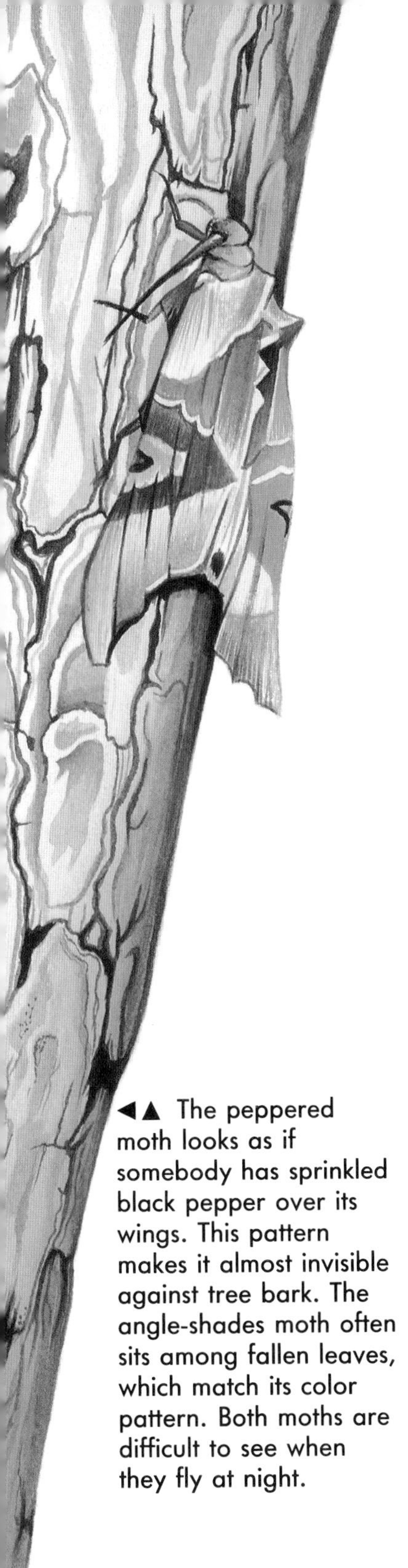

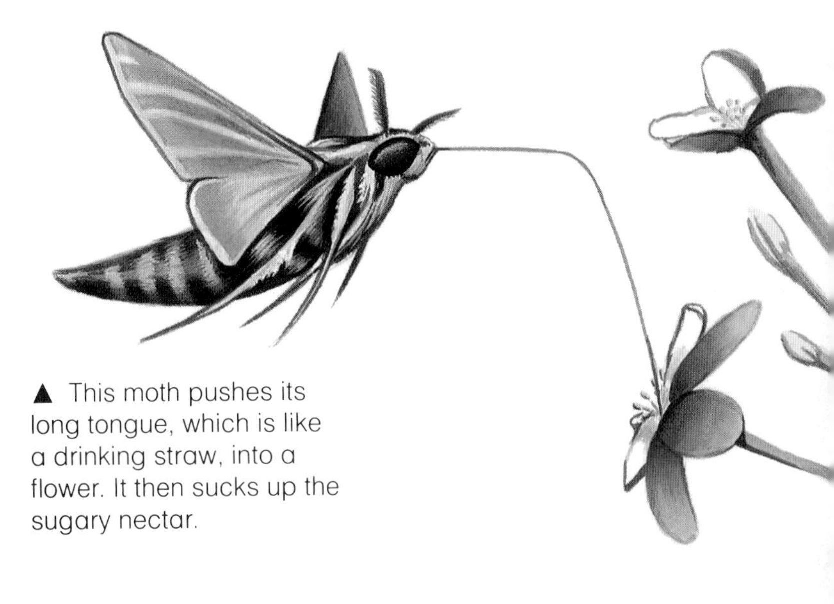

▲ This moth pushes its long tongue, which is like a drinking straw, into a flower. It then sucks up the sugary nectar.

Moths have large eyes, like butterflies, but it is unlikely that they can see bright colors. The flowers that they feed on are mostly pale, so food would be easy to find on all but the darkest nights. The flowers all have a very strong scent, which probably guides the moth to the food area.

Male moths also use their sense of smell to find mates. Female moths give off tiny amounts of scent that carry on the wind. Males fly into the air currents, using their huge *antennae* to follow the odor. A male moth is known to have tracked a female across a city, ignoring all smells but the one that she produced.

◄▲ The peppered moth looks as if somebody has sprinkled black pepper over its wings. This pattern makes it almost invisible against tree bark. The angle-shades moth often sits among fallen leaves, which match its color pattern. Both moths are difficult to see when they fly at night.

Animals That Light Up at Night

After dark, people often use flashlights. Many small animals become living flashlights at night, but they use their light for signaling, not for finding their way.

Animals make light by mixing special chemicals in their cells. The lights that we use always make heat, which is wasted. Animals' chemical light is cold, and wastes very little energy. Unfortunately, nobody has found out how to make these chemicals on a large enough scale to make light for us.

▼▶ Glowworms and fireflies make the brightest lights of all land animals. You might think that having a light would make them easy prey, but they are quick to switch off if they sense any kind of danger.

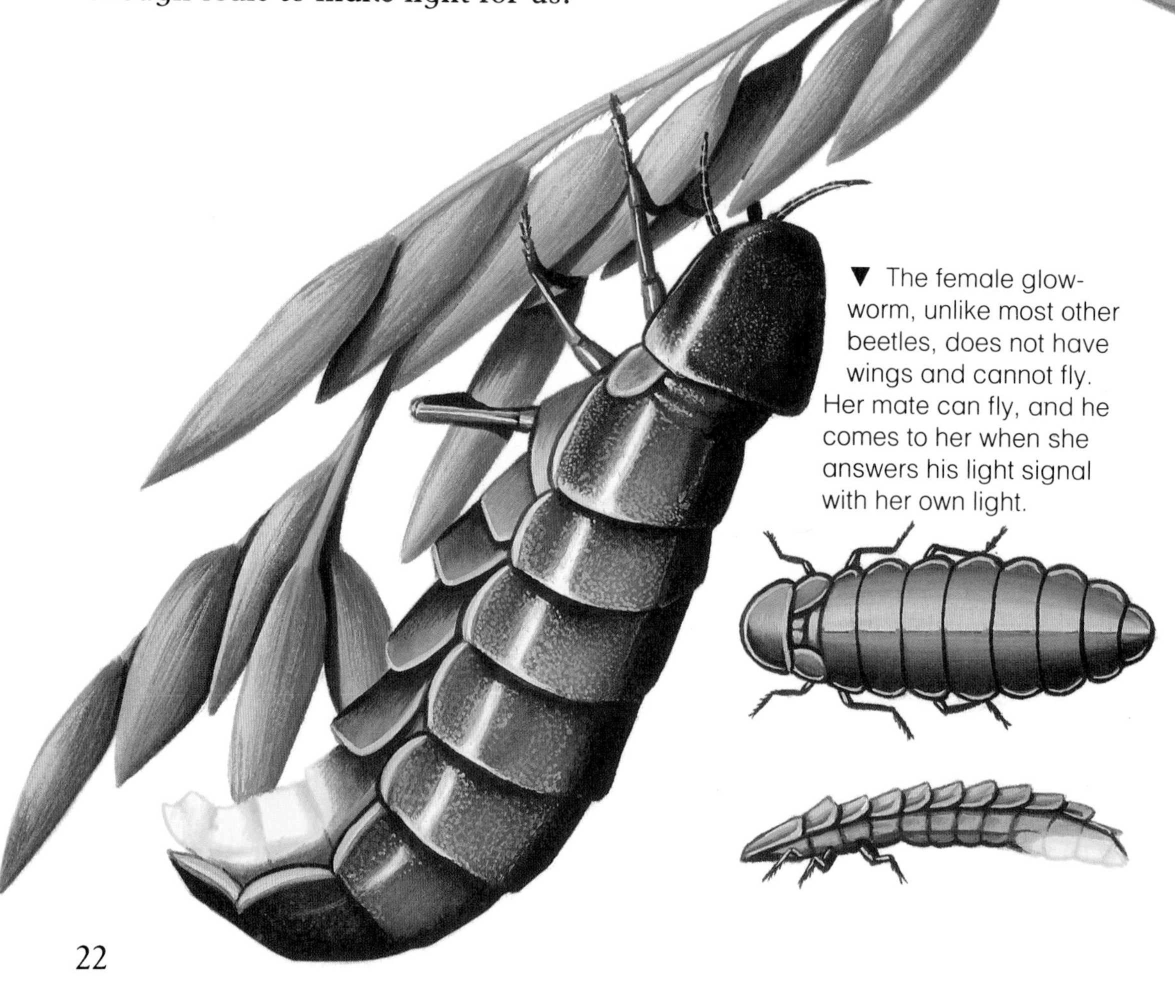

▼ The female glow-worm, unlike most other beetles, does not have wings and cannot fly. Her mate can fly, and he comes to her when she answers his light signal with her own light.

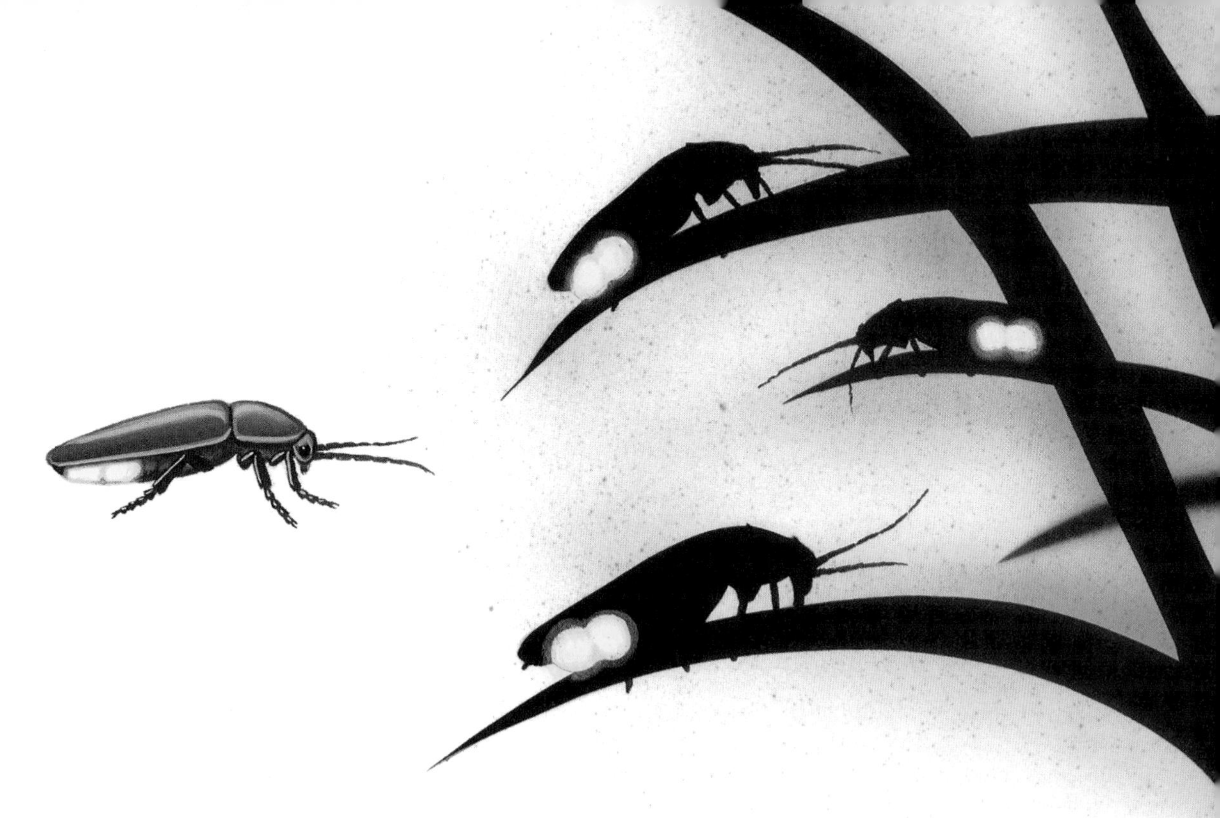

In a few animals, light is produced throughout the body. In most, it is concentrated in a few areas. In the deep sea, where it is always dark, some fish and squid have light organs with lenses that concentrate the light, like old-fashioned lanterns. Land animals do not have such complex arrangements, though the brightest of them would give enough light to read this book by night.

Fireflies and glowworms are beetles that produce light. Some females are wingless, and signal with their lights to males, which can fly. In other kinds of fireflies, males and females both fly. Sometimes they signal by flashing their lights on and off in a kind of insect Morse code. In some parts of the tropics, this signaling may go on for hours every night.

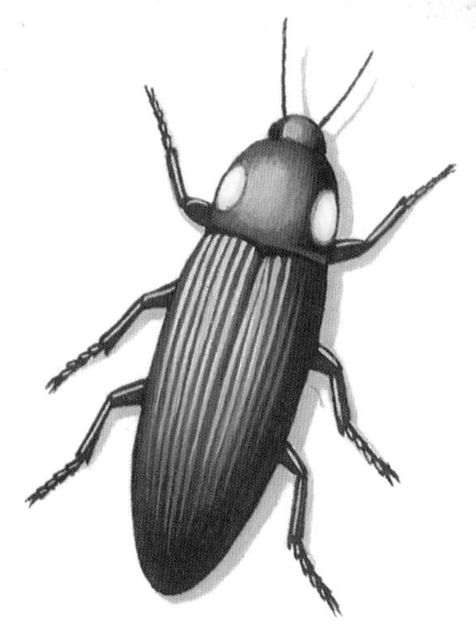

▲ The brightest of all lights is made by an American beetle called *Pyrophorus*.

Reptiles at Night

When the sun sets, the Earth cools, and as the temperature drops, dew may form. The cool, moist night suits many kinds of animals that cannot stand too much heat. In deserts, which can look lifeless during the daytime, insects, reptiles and mammals leave their burrows and look for food in the cool of the night. Even in wetter parts of the world, slugs, earthworms, and wood lice feed at night, because they dry out quickly if they are exposed to the heat of the sun.

Cold-blooded animals, such as reptiles and amphibians, often come out at night. Most of them are small, and in hot climates they can die of heat stroke during the daytime. They control their body heat by moving from hot to cool places.

▼ During the breeding season, male frogs sing to their mates at night. Some frogs croak, others make trilling or whistling noises.

► When dusk falls, geckos come out from their hiding places to feed on insects. Their large eyes enable them to see their prey, even when there is little light.

Many of them, from frogs to crocodiles, court their mates after dark. This is probably because the calls they make could attract enemies. At night they are more difficult to find. Even so, the frog calls break off immediately if there is a strange noise.

One reptile, the tuatara, lives on small islands off the coast of New Zealand, where the climate is cool and moist. It leaves its burrow after dark to forage for insects. The temperature may be as low as 50°F (10°C). Most other reptiles are not active when it is this cold. Nobody knows how the tuatara copes.

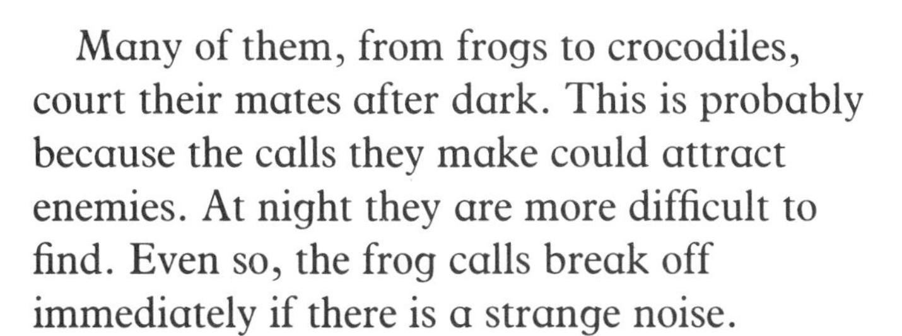

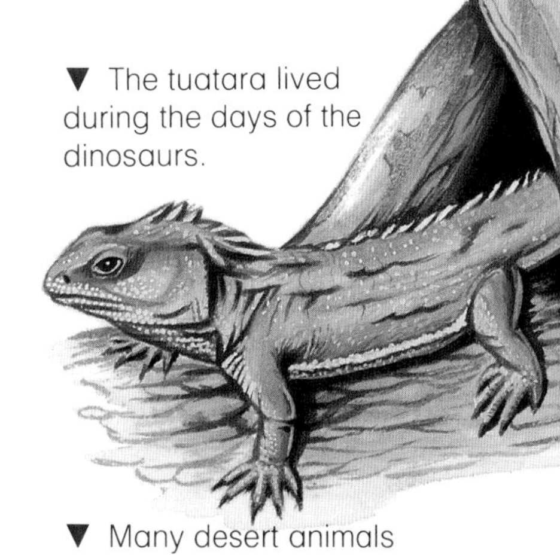

▼ The tuatara lived during the days of the dinosaurs.

▼ Many desert animals look for food at night. Jerboas search for seeds and plants. They have to watch out for predators such as snakes. Scorpions may eat small mammals, but usually feed on insects.

The Sea at Night

The sun's rays do not reach far into the sea, and most ocean waters are always dark. In spite of this, many creatures of the deep sea lead a life that follows a daily pattern. Some of them come to the surface at night, perhaps attracted by the huge numbers of tiny plants that float there. They are followed by fish and squid, which are often the prey of fishermen. As they stir the surface of the water, some of the tiny plants give off flashes of light. Nobody knows why they do this, but in warm parts of the world, the sea often blazes with living light.

▼ An octopus leaves its lair at night to hunt for crabs and other creatures. The octopus is difficult to see, but its enemy, the moray eel, can smell it. If the octopus is attacked, it squirts out ink as a decoy.

On the beach, animal activity is controlled more by the tides than by light. On moonlit nights when the tide is out, wading birds may feed on mud flats. But creatures such as barnacles and mussels, which have no eyes, feed whenever the tide is high.

Some offshore animals become active at night, and one of the best places to see this is a coral reef. As the sun goes down, many of the corals spread their tentacles to feed. Most of the brightly colored fish disappear to rest in secure hiding places. Their place is taken by starfish and sea urchins, and their strange relatives, the feather stars and basket stars. Octopuses emerge from their lairs to hunt crabs and lobsters, and are themselves hunted by the stealthy moray eel.

► Squid hunt for small prey at night. They often catch insects that fall into the water.

The Hunted

Many nocturnal creatures are hunters. They move quietly to catch their prey, which they detect with their ears and their sense of smell. The creatures that are hunted must also have good senses, and the ability to escape quickly.

Mice come out to feed at night. Their huge eyes, ears, and whiskers warn them of enemies such as cats, owls, and snakes. They can leap out of danger. Some have tails that break easily. A predator may find that all it has caught is a tiny piece of skin.

▼ This stoat just missed a meal. It caught the tip of the tail of a wood mouse, but the skin broke and the mouse leapt away to safety. In a day or so, the tip of the tail will dry up and fall off. This is why you sometimes find a stump-tailed mouse.

Many insects are nocturnal. Ground beetles are hunted by shrews and hedgehogs. Some protect themselves with nasty-tasting chemicals. In the air, bats prey on moths. Some moths have ears that can hear a bat's echolocation squeaks. They then take evasive action by tumbling into a different line of flight. Others escape because they are able to make tiny sounds that confuse the bat's sonar.

On a dark night, some worms leave their burrows to find food. They are among the most cautious creatures. If you go out with a flashlight, you can see the worms shoot back to safety underground when they feel the earth tremble from your movements and detect the light. All hunted animals have ways of escaping. Some are caught, but neither hunters nor hunted win all the time.

▼ Some kinds of worms leave their holes at night to look for leaves, which they pull underground and eat. Though they cannot see, they are wary and hide quickly if they sense any movement nearby. In spite of this, they are caught by their predators, which include badgers and owls.

► Slugs come out at night to search for food. During the day they need to hide in cool, moist places, otherwise their soft bodies would dry up.

Glossary

antennae the pair of feelers that insects and some other small animals have on their heads. The main function of antennae is probably to smell, rather than touch.

cold-blooded having a body temperature that changes with the environment. Amphibians and reptiles are cold-blooded. When it is cold, their body temperature drops. When it is hot, their body temperature rises and may be much higher than that of any warm-blooded animals. The advantage of being cold-blooded is a low energy system that needs little food. The disadvantage is that cold-blooded animals have to protect themselves from intense cold or heat, which they cannot survive.

diurnal active during the daytime.

echolocation the ability to calculate how far away an object is by bouncing echoes off of it. Animals such as owls and bats are able to find their way in the dark and catch food using echolocation.

environment an animal's living area and all that it contains.

migration a regular journey, usually every year, of part or all of a group of animals from one part of the world to another, to avoid cold weather and to find a plentiful supply of food. Some journeys are long while others are quite short. Many animals migrate each year, including insects, fish, birds, and mammals.

nocturnal active during the night.

sonar a kind of echolocation that uses only sound waves.

torpid a state of temporary hibernation in some animals. Animals become torpid to reduce their energy levels when it is extremely cold or when they cannot find enough food.

vespers flight the evening flight of some birds and bats.

vibrissae the long hairs around the snout on some mammals' faces. At the base of each vibrissa is a large nerve ending. This makes the hair sensitive to touch.

Index